Building Stone Walls

Second Edition

John Vivian

Storey Books
Schoolhouse Road
Pownal, Vermont 05261

The mission of Storey Communications is to serve our customers by publishing practical information that encourages personal independence in harmony with the environment.

Copyright © 1976, 1978 by Storey Communications, Inc.

Storey Publishing books are available for special premium and promotional uses and for customized editions. For further information, please call the Custom Publishing Department at 1-800-793-9396.

Printed in the United States by Capital City Press
Illustrations by Douglas Merrilees and Ralph Scott

30   29   28

Library of Congress Cataloging-in-Publication Data
Vivian, John.
    Building stone walls.
    1. Stone walls.   I. Title.
TH2249.V58   693.1   75-20773
ISBN 0-88266-074-8 pbk.

Garden Way Publishing was founded in 1973 as part of the Garden Way Incorporated Group of Companies, dedicated to bringing gardening information and equipment to as many people as possible. Today the name "Garden Way Publishing" is licensed to Storey Communications in Pownal, Vermont. For a complete list of Garden Way Publishing titles call 1-800-441-5700. Garden Way Incorporated manufactures products in Troy, New York, under the TROY-BILT® brand including garden tillers, chipper/shredders, mulching mowers, sicklebar mowers and tractors. For product information on any Garden Way Incorporated product, please call 1-800-345-4454.

Contents

INTRODUCTION 1

KINDS OF STONE 4

EQUIPMENT 16

LAYING OUT THE WALL 22

ABOVE GROUND BUILDING 29

ENDS & CORNERS 36

USING NOT-SO-GOOD STONE 45

DRAINAGE 52

SPECIAL WALLS 54

WALL FURNITURE 57

QUARRYING 66

MORTAR 72

MAINTENANCE 76

SOME OTHER USES OF STONE 78

THE STORY OF A WALL 96

"There is
no such thing as
a half-built stone wall.
It's either a wall
or a stone pile."

Introduction

If you stop and think about it it seems that, written word aside, the most enduring monuments to man's creativity and hard work are built of stone. The Pyramids in Egypt, the Great Walls of China and Peru, temples most everywhere from Latin America to India, the castles of Europe and the mile on mile of stone walls running through our New England countryside were all laid up by hand and without a speck of mortar.

They endure in part because rock is as near a definition of "forever" as exists. *The Pull of Gravity* But more important is their main construction ingredient—gravity. In a properly-built stone wall each rock sits square on the ones below it, and so long as gravity keeps pulling on every stone in it, that wall is going to stay put.

So, here's how to go out and build a really permanent monument to yourself. Do it right, square and plumb and well-tied throughout, and the wall will be standing long after you and I and all our other accomplishments and failings are forgotten.

1

You'll Need
Boots First thing, get a pair of well fitting steel-toed boots with ankle, or better yet, calf-length uppers made from stiff, thick leather. Next, buy a pair of horsehide work gloves or a pair molded of rubber with grit imbedded into palm and fingers. These are the better to grip stones with and to keep your toes from getting dented when you drop a boulder on your foot—which you are bound to do at one time or another. Then, you'd best be sure you have the needed time and ambition.

There is no such thing as a half-built stone wall. It's either a wall or a stone pile. And to get from one to the other takes a lot of lifting. A cubic foot of rock weighs the better part of a hundred pounds.

Heavy Work So a little decorative wall only three feet high, two feet wide and twenty feet long weighs some five tons or more, depending on the amount of airspace built in, and it comprises a thousand or more average-sized stones. If you have to fetch stones from somewhere, there is the loading and unloading in addition to the building to consider. That little twenty-foot wall can have a man lifting well over twenty tons of dead weight before it is finished.

If you are old enough to remember listening to *Fibber McGee and Molly* or *Your Hit Parade*—before television—I strongly suggest you get your physician's OK before taking on a stone wall, especially if you are a desk worker and unused to strenuous labor.

In any event, plan to take your time and use carts, ramps, barrows and levers to move larger stones. There's little point hurrying to complete a wall that will likely endure into the next millennium. And no point at all in busting a gusset doing it.

Kinds of Stone

The next step is finding stone. Most everything I've read about building with stone assumes you have a sufficient supply of "good," easy-to-build-with stones with three or four flat sides. And they suggest that if you don't have enough good stone lying around you can buy or quarry it or dress rounded stones flat.

Use What's There Quarrying by hand and dressing stone is devilishly hard work, and paying to have rock trucked any distance is expensive. Certainly, good stones are easier to build with and make a classier looking wall. But I'd say, if some or all of the stones you have are in the "not-so-good" category, go ahead and plan to use them. We'll get into using both kinds for wall construction.

In New England, much of the Midwest, the Lake States, and parts of the Northwest, the ancient glaciers either deposited or exposed many types of stone suitable for building purposes.

The glacial deposits, composed of sand

4

and gravel, form the *till plains, moraines, eskers,* and *drumlins.* In addition, there are huge boulders, often referred to as *glacial erratics,* which litter the fields, forests, and streambeds—all giving the glaciated countryside of these regions their topographical character and form.

In some places, the mile-thick, ancient ice sheet had scoured, gouged, and literally plowed away the earth's mantle; leaving the area's bedrock exposed in its retreating wake. Today, one can find many different kinds of stone for building purposes. There are three basic types—*igneous, sedimentary,* and *metamorphic.*

Three Types of Stone

Glacial advance during Pleistocene period of continental ice sheet.

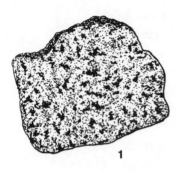

1

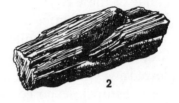

2

3

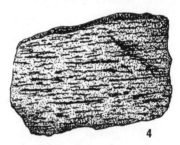

4

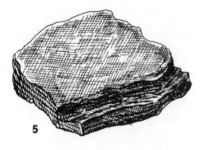

5

Rock types:
1. Granite (Igneous rock)
2. Shale (Sedimentary rock)
3. Limestone with fossils
    (Sedimentary rock)
4. Gneiss (Metamorphic rock)
5. Slate (Metamorphic rock)

Igneous rocks result from the cooling and solidification of hot liquid rock. Much of it was formed deep in the earth, and it is the hardest, densest kind of rock to be found. An example is *granite*—the most common *intrusive* igneous rock, meaning it was cooled and solidified at great depth. This is opposed to *extrusive* igneous rock, such as *basalt,* which cooled and solidified at the earth's surface.

Sedimentary rock is produced by the consolidation of sediments that have been formed by weathering and piled up by any of the agents of erosion (wind, water). Thus, the most noticeable feature of sedimentary rocks is their layering. However, many sedimentary rocks preserve features or objects associated with the sediments, such as shells, bones, invertebrate animals, plants, footprints, ripple marks, and mud cracks. In other words, sedimentary rocks often are *fossilized,* and include such common stones as *sandstone, limestone, shale,* and *conglomerate.*

Metamorphic rocks are the result of either sedimentary or igneous types being modified by pressure, heat, and chemically active solutions. Such rocks are characteristically banded or layered in some fashion. Some common metamorphic rocks are *marble, slate, schist,* and *gneiss.* Marble is metamorphosed limestone, slate was originally shale, and schist and gneiss are modified granite.

7

Quarry waste of metamorphic rock provides chunky but good material, in this example poorly laid up.

Some Will Split   Probably the best wall stones come from hard shales and schists—rock types that developed flat cleavage planes during metamorphism so they split out into layers with flat tops and bottoms. Many will break naturally into stones with flat sides as well. Others may not come with good flat faces, but can be dressed quite easily. I've worked with some metamorphics that split almost as easy as a block of wood.

Hardest to build with are igneous stones found in fields or running water. Glacial action or gradual erosion in a stream has tended to round even those very dense and hard rocks. Though nearly all rock has a "grain," or a tendency to split along a fair-

8

ly flat place, finding the grain in granite and the like is difficult, and splitting faces off small rocks takes more time and effort than it is worth.

Stone mason Ted Metevier splitting stone for driveway edging wall. Note curved retaining wall in background.

Heavy and irregular stones for base course are dug into ground.

Laying out the first wall courses while stone is being sorted.

10

Sedimentary rock formations, having been laid down in sheets, tend to occur in layers—properly called *strata*. Most are relatively soft and easily split or cut by nature or a quarrier into good building stone. But they also wear faster than other kinds of rock. Wind-borne particles will wear down sandstone in time, and limestone gradually gets eaten away by the natural acids in rain water. A sandstone or limestone wall might not hold up for more than ten or twelve thousand years.

Besides its good breaking tendencies, metamorphic rock is usually durable—even though it is often soft and easily worked. Marble is the traditional medium of sculptors and much ancient Greek and Egyptian marble sculpture has withstood wind and rain for thousands of years. Another metamorphic, slate, was used for gravestones by the earliest settlers in America. Go to an old colonial graveyard and notice it is the slate stones whose inscriptions are still sharp and clear. Other markers, igneous rock included, have worn nearly flat.

Other Materials   In some areas, of course there is little or
no exposed rock. In many of the Plains
states you will have to go to a river valley
or excavated quarry to find stone. In the
Florida Keys I've seen dandy walls made
from coral chunks. And in parts of the
Southwest folks have to make their own—
soil, straw and water packed in molds and
dried in the sun to adobe. It all makes a
good wall.

Sources of Stone   Lacking a supply of stones on your own
place and not wanting to buy them, you can
look several places. Perhaps another land-
owner will let you haul off rocks from his
walls or abandoned stone buildings. Rock
ramps, cellars, and foundations left after an
old house or barn has burned down provide
one of the best "good" stone sources you
can find. Often these old cellar holes are
dangerous or an eyesore, and owners are
glad to have part of the demolition or fill-in
job done free.

12

Construction sites often provide excellent flat-sided rock picking, especially where new highways are being dynamited through the hilly country. Streams, rivers, many lakes and the seashore in some places are good sources for frost-split or water-rounded stones. And you'll find that stones only weigh half as much moved under water.

If there are a good many old stone buildings or foundations in your area, you may find abandoned quarries or gravel pits scattered through the countryside that were dug by the original stone-builders. Ask around, or consult a U.S. Geological Survey topographical map. Abandoned and active quarries and gravel pits are shown on such detailed maps. *Quarries and Pits*

If you are unfamiliar with *topo maps*, they are a must for the stone-builder in search of material. A topo map is essentially a detailed rendering of an area covering 49 to 70 square miles called a *quadrangle*. The map shows roads, buildings, streams, lakes or ponds, swamps, contours, and elevations (in 10-foot intervals), as well as quarries and gravel pits. *Look on Maps*

Each quadrangle is named for a town or important feature within the area covered, and each state is divided into numerous "quads." To find out which quadrangle you need, you must first obtain an *index*, which is a map of the state showing the quad divisions and their names. To order topo maps and a state index, write to:

U.S. Geological Survey
1200 Eads Street
Arlington, Virginia 22202

if your state is *east* of the Mississippi River. If you are *west* of the Mississippi River, then address your inquiry to:

U.S. Geological Survey
Federal Center
Denver, Colorado 80225

Tell About Stones
Also, some quadrangles have been mapped geologically. That is, there are special issues of certain quads which have been colored to show glacial and/or bedrock features of the 49- to 70-square mile area. It is worthwhile to inquire about such quadrangles because they give detailed information on the type of stone you are likely to find, where large boulders are likely to occur (if a glaciated area), and where bedrock is exposed. Not all quads have been so mapped, however.

On a broader scale, there are some states which have maps showing its geological features. The State of Ohio, for example, has a state map of its bedrock geology and another of its glacial features. You should check with your *state geological survey* to find out if such maps are available for your state. If there is no such governmental body in your area, then inquire about these maps to the U.S. Geological Survey at the addresses previously given.

Finally, you may have to buy stone or re- sort to digging, quarrying, or picking up your own from fields and creek beds as nature provides them. Operating quarries are often listed in the yellow pages of your telephone directory. In addition to map sources and the state geological survey, you may have a geology department of a college or university nearby that can give you further information, especially on what kind of rock is available and how to obtain permission to quarry or dig your own stone.

We will go briefly into quarrying later on; for now, let's assume you have your stones at hand. The principles of wall building are pretty much the same no matter what kind of rocks you use. Let's build a section of that basic three-by-two-foot wall to get the essentials down pat.

Equipment

Cart Is Useful I've found one of the handiest pieces of wall-building equipment—better than a wheel barrow—to be one of those high-wheeled, box-shaped garden carts. You can tip one up on its square end and just roll the stones in. Don't overload it, as I did once, and have a wheel collapse on you. Used sensibly, the carts save much lifting, and you can run larger stones up planks to the wall top with one.

Other tools include a yardstick, a hank of stout cord, several stakes a foot or so longer than the wall is to be high, a line level to go on the cord, a mason's level and perhaps a long and true board to tie it to, plus a good-digging spade.

For the Big Ones To move larger rocks around, you'll need a long (five feet or more) steel crowbar, and a smaller crowbar, or *pinch bar*, that has one end curved. Another device is the *hoe/pic*. This tool is 26 inches long, weighs about 4 pounds, and is somewhat expensive (about $17.00). However, it has many uses—pick,

16

Using Garden Way cart for loading & hauling stones.

wedge, prybar, and a hoe. Its versatility goes beyond the use in stone wall building and it has a decided advantage over conventional prybars when one must dig large boulders that are embedded in the ground. You'll also need a collection of thick planks and short lengths of $1^1/_2$ or 2-inch iron pipe to serve as rollers if you are working with really big stones.

If you plan to do any trimming or dressing of stones, get a pair of safety goggles. Rock can splinter into razor-sharp fragments, and even a dull chunk in an eye can mean trouble. Be sure the goggles are well ventilated so they won't fog up on you during a hot day's work. *Goggles Needed*

For trimming and dressing you should have a set of mason's or geologist's ham-

Wall-building tools.

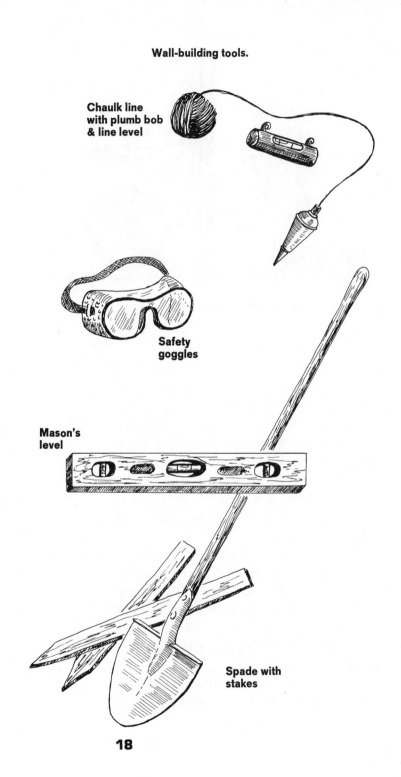

Chalk line
with plumb bob
& line level

Safety
goggles

Mason's
level

Spade with
stakes

18

mers and chisels. These come in a variety of widths and shapes, and are used mainly for scoring and splitting both brick and stone. A relatively new tool available to stone-builders is the rockhound's *gad-pry bar* (cost about $7.00). Used with a heavy crack hammer, this 18-inch tool easily opens seams and crevices in stone. It has two hammering faces (see tool illustration) that permit you to drive the bar down into a seam, and then to drive it to the side to force the crack or seam apart.

Handy Tool

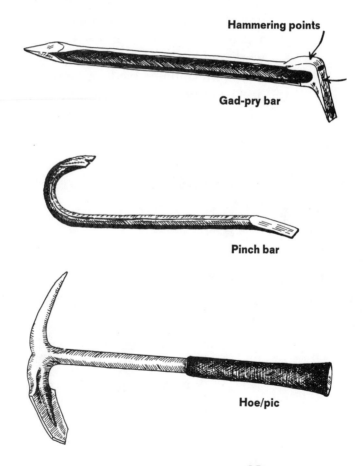

Hammering points

Gad-pry bar

Pinch bar

Hoe/pic

Rock Chisel   If you plan to do any drilling or carving in the rock, an artist's supply store can offer you a selection of rock sculptor's tools. For drilling, you'll need a high-tempered rock chisel with a cutting head $3/4$ to $1/2$-inch wide. It will be forged in a dull "V" shape and will serve for carving, too.

Types of Hammers   Hammers come in two types. One kind has one flat head and the other in a wedge shape—this is the traditional stone mason's hammer. The other, a Bush hammer, has a flat, toothed head for really getting a purchase on a piece of rock. You can also get a lightweight mason's hammer shaped like a geologist's pick. It is for more delicate work and has a small pounding face and a long, thin chipping blade on the other side. Special heavy-duty tools will be mentioned when we get to quarrying.

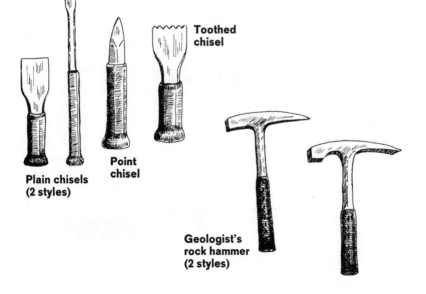

Toothed chisel

Plain chisels (2 styles)

Point chisel

Geologist's rock hammer (2 styles)

Pounding or crack hammer

Stone mason's hammer

Bush hammer

Most tools for stone-building can be purchased at hardware stores and building supply houses. Some, however, are rather specialized geological tools obtainable only at certain suppliers. To obtain such items as a geologist's hammer, hoe/pic, and gad-pry bar, write for a catalog from the following suppliers:

Forestry Suppliers Inc.
205 West Rankin Street/Box 8397
Jackson, MS 39204

and

The Ben Meadows Company
3589 Broad Street
Atlanta (Chamblee), GA 30366

Laying out the Wall

Check Property Line If the wall is to run along your property line, be sure that the whole thing is within your boundaries, unless your neighbor is eager to share in cost, construction and upkeep of a shared wall.

In the old days New England farmers would patrol shared stone fences each spring, each replacing the smaller winter-dislodged stones on his own side, the two of them joining forces on the big ones.

As the crochety Yankee farmer in Robert Frost's *Building Wall* puts it, "Good fences make good neighbors." Don't know as I agree with the sentiment. Frost didn't. I'd say, don't build a shared fence astride your property line unless you are *already* good neighbors and plan to remain so for the next several generations.

Plan It First thing, lay out the outline of full length and width of the wall with cord looped to short sticks. For a curved wall you may want to lay out with thick rope or garden hose to describe a fair curve.

22

If you can't avoid them, grub out any trees, stumps, or underbrush in the wall's area. Remember, that little maple sapling a few feet from the wall is going to grow. In time its roots will heave the rock, and the trunk may expand and push the wall aside. Cut it down.

Clear the Way

A low wall can be built right on the ground. In a few years the lowest stones will sink into the sod a bit and no one will know the difference. It is better stonemasonry, though, and makes for a sturdier wall, to dig out sod and topsoil so the footing—the lower courses or layers of rock—rests on the underlying subsoil or hardpan.

May Need Footing

Most places where you find plenty of na-

This wall has stood for generations.

23

tive rock the topsoil layer will be shallow, a foot or less deep, and removing it will be a minor chore, likely turning up an additional supply of rocks in the bargain. In some valley and lowland areas with deep, loamy topsoil a stone wall just set on the surface would gradually sink out of sight. But then, I don't think an honest stone wall would be comfortable in the flatlands.

Square It Up   Footing trench or no, the next step is to lay out your batter boards, stout stakes hammered in to mark the four end corners of the wall. Put the stakes in good and deep and use your mason's level to make them plumb—straight up in all dimensions. Make sure the tops of the stakes are several inches higher than the planned wall.

Cord Is Guide   Next, tie your cord to the four stakes at wall-top height and stretch the cord as tight as you can. Every six to ten feet, on both sides of the wall, hammer in more stakes. Make sure they are outside the wall area so string touches the inner-facing sides of each stake. You may tie or staple line to the stakes if building up or down grade or if the wind is bellying the upwind line on you. (A good many wall-builders, this writer included, have neglected to use the auxiliary stakes only to find that their supposedly straight wall ended up with a slight curve due to the prevailing wind blowing out the guidelines.)

Now attach the line level to the cord and adjust till all four of the sides and both ends are level. This will define the approximate plane of the top of your wall. For most pleasing appearance, sturdiest construction and most satisfaction from the work, the wall top should be flat and level from side to side, either following the lay of the land in the long dimension or remaining horizontal, taking grade in graduated steps.

On a grade or flat, courses should be about the same thickness, each course running horizontally—parallel to the level. Sides of the wall should be vertical, or in higher walls they should have a slight inward slope (a slight batter) on each side. Ends and corners should be square and vertical.

Level line across the top of a retaining wall proves it's true.

25

Keep It Plumb   This isn't feasible 100 percent of the time with stone, but try to keep all dimensions as plumb and square as you can. You'll be looking at that wall probably for the rest of your life, and come fire, hail or high water it will be one thing standing for your great-grandkids to remember you by. Leave them the best job of work you can.

Also remember that gravity pulls straight down. Unless the wall rests on a flat, horizontal and level plane (or sections rest on a succession of flat steps on hilly ground), gravity will slowly pull your wall downhill. So, either level out the ground or dig footings with bottoms having a plane parallel to the guideline. See the illustrations in *Footings* for several ways to achieve this.

A lot of books tell you to put your

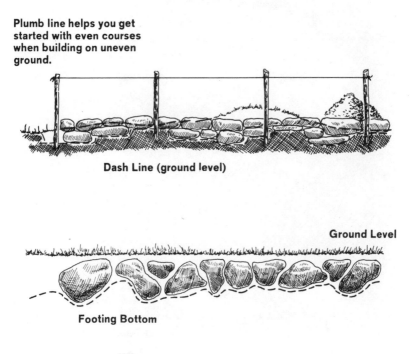

Plumb line helps you get started with even courses when building on uneven ground.

Dash Line (ground level)

Ground Level

Footing Bottom

On steep grade base course stones are dug into steps.

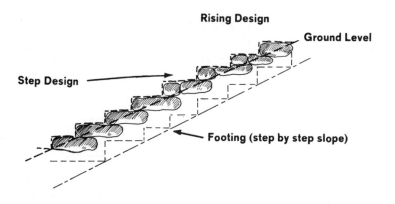

Rising Design

Ground Level

Step Design

Footing (step by step slope)

biggest, flattest shape stones at the bottom of the footing—then later on tell you to save them for the topping course. Having worked mainly with odd-shaped stones the glaciers left in our corn fields, I pick out the absolute worst stones, the ones with not a single flat surface or with odd-shaped protuberances. I find the least unreasonable side, then bury the stone in whatever shape hole is needed to get the best side exposed at the depth I want the bottom of the first course to run.

Worst Stones First

The objective in all this is to give the wall a good level base to rest on. Even if you must dig a series of notches in a hillside, your wall will be the better for it. In all below-grade work, keep stones several inches apart, filling the open space with smaller rocks. This is to permit water to drain through easily.

Give it a Good Base

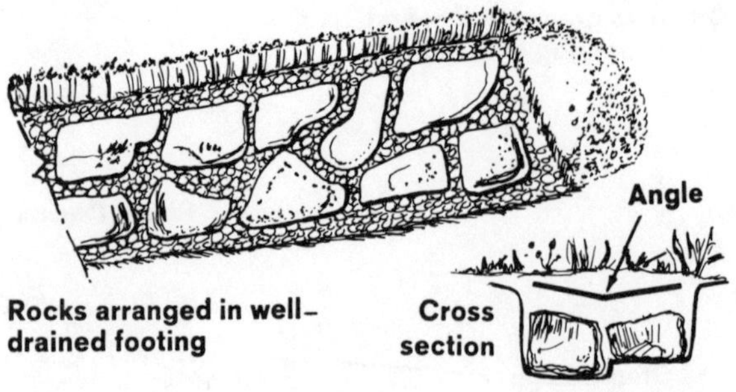

Rocks arranged in well–drained footing

Cross section

Angle

Slight Lean   The footing course or courses should be laid to be a bit lower in the center—higher out at the edges. This slight "V" angle is often maintained throughout construction. Thus the outer walls of the structure lean in against themselves. Gravity helps keep the wall standing by pulling rocks down as well as "in" toward the wall's center.

Above Ground Building

With the footing laid to ground level, lower *Save Lifting*
the guidelines to what will be the top of the
first above-ground course. This should be
the average height of the thickest, heaviest
rocks you have. No point lifting them any
higher than necessary.

Do try to save the flattest rock with most
uniform thickness for the top. The bigger
the better, though don't save out any stones,
no matter how flat they are, so large you
can't handle them easily at the top.

Now begin laying wall. Keep the best flat *Best Sides Out*
face of the narrow dimensions of each stone
facing out when possible. Be sure each stone
is bedded solidly on the stones below it. If
a stone wobbles it is better to chip off a
wobble knob or dig out a hole or make a
joint in the rock below than try to shim
it up with small rocks and wedges. If you
do use smaller rocks to get the wobble out
of a big stone, be sure they are wedged in
tight and held in place by other large rocks.

Keep the guidelines level, and continual-
ly sight along the side of the course, adjust-

29

Wobble knob

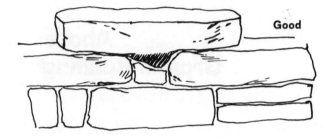

Good

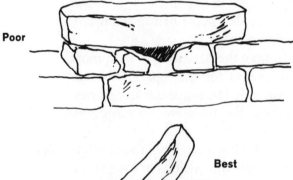

Poor

Best

ing rock placement with level and yardstick to make sure all continues square.

To be sure the sides are as vertical as possible, or that they slope inward at the desired angle, hang a small pebble from a length of string.

Put a bent wire on the other end and hang *Use Plumbob* it from the guide line. Run the plumbob along as you lay up wall. Keep moving the guidelines up as you finish each course.

With a small wall only two feet wide and with reasonable luck in getting good stones, your courses seldom will be more than two stones wide, and a good many of the bigger stones will reach all the way from side to side.

Don't use any small stones in the outer *Small Stones* faces; they'll be the most likely to work *Inside* loose in time. Put the little stuff in the interior to fill gaps between the larger stones.

Often you'll find a good-fitting stone that is too big; one end or corner sticks out. Then, take your wide chisel, score all around the chunk you want to remove— score $1/2$ to $1/8$-inch deep all around if you can. Then knock off the extra. You may have to get some heavy-duty equipment such as a heavy sledge. If the steel strikes sparks you probably have an igneous rock and should find another place for it to fit unless you want to spend a long time hammering.

With good, longish, relatively flat stones, *One on Two* you should never have a vertical fissure in the wall that extends up from course to course. In other words, each rock should rest on at least two others in the course below, and joints between stones should not extend from course to course. Too many

31

such adjacent joints is called "run" and it greatly decreases your wall's stability.

It Will Move   You may not see it move, but any dry-laid (mortarless) wall is in continuous motion during settling, as soil is moved around underneath by flowing water and, in the North, by alternate freezing and thawing of the earth.

With too much run, the stresses will concentrate at the weakest point, where a joint extends through two or more courses. In effect, the wall will try to fold there, and in time it will fall out.

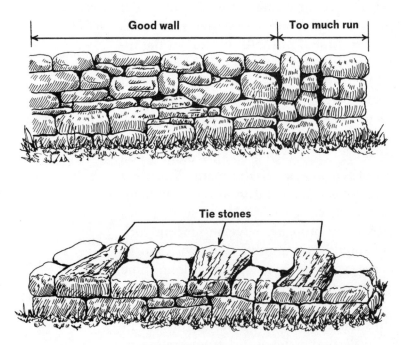

Good wall | Too much run

Tie stones

Gophers, squirrels, woodchucks and all manner of other wildlife find a stone wall a natural home site. The wiggling of a full-grown groundhog or a litter of rabbits can

Looking over the possibilities and grain of a thick stone.

Chinking in is the way to do it here.

33

A ragged stone end is chipped straight atop an angle iron. The stone boat makes an improvised work bench.

A topping stone is placed on a corner, carefully pre-built to accommodate the stone's slight arch.

A nicely laid section of wall crosses a ditch evenly.

dislodge a stone that is too small or poorly set. So can a dog digging after a chipmunk or the vibrations as youngsters or hunters wander along the top of the wall.

From time to time—every six to eight feet if you can—place a long stone with its longest axis aiming into, rather than along the wall. Alternate sides in doing this; it is called "tying" the wall. The tie stones keep the wall from consisting of two unattached outer layers; the tie stones literally tie one face to the other.

Tie the Wall

Ends & Corners

Best Rocks at Ends Save the best rocks, those most nearly rectangular and of even thickness, for ends and corners. A wall end must stand on its own, and for appearance's sake the end stones should have at least one good square corner —or those at the outside edges should. Top stones should be the biggest you can handle. I like to keep the longest rocks for ends, so the end rocks will extend as far back into the wall as possible. It is also a good idea to have ends particularly well tied crosswise.

Turning a Corner Corners are more complicated still. First, you want to tie your corner into both lengths of wall—which is to say that in each course it is well to have an extra-long rock extending from the corner into both the north/ south and the east/west length of wall. If you lack enough long stones to tie into both walls' each course, alternate them—tying first into one, then the other.

From the footing course up, you must take another precaution. Remember that the corner must withstand expansion and contraction from both lengths of wall. A bias toward the layup pattern of one length over the other would cause the corner to work loose in time.

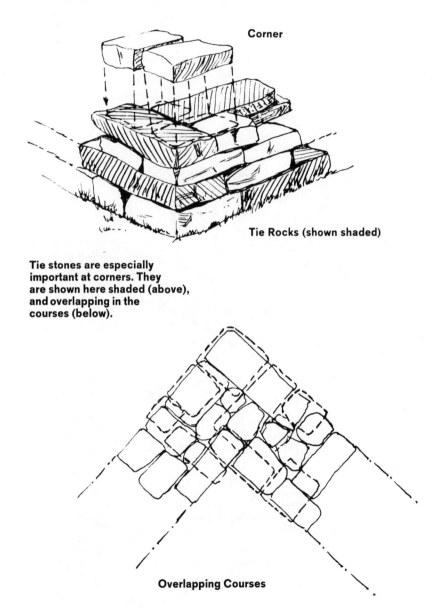

Corner

Tie Rocks (shown shaded)

Tie stones are especially important at corners. They are shown here shaded (above), and overlapping in the courses (below).

Overlapping Courses

So try not to let any edge of any rock line up with any *joint* in the course below. Or, don't let the edge of any rock line up with the *edge* of any rock below—except, of course, on the outer and inner faces.

Good End (tie stones shaded)

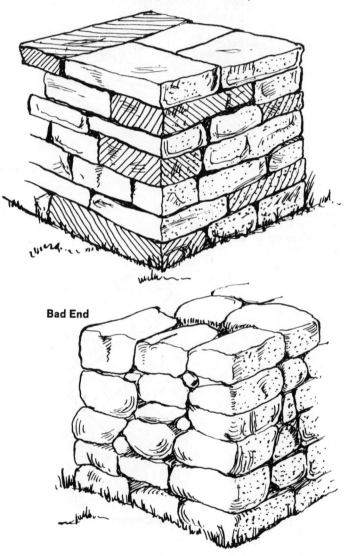

Bad End

Roughly-dressed stone provides the firm foundation for this old mill.

In still other terms (this being a bit complicated to explain) put each stone so as to cover as many joints as possible, being sure each covers some north/south joints and some east/west joints. If this still doesn't make sense, the illustration speaks better than words.

So now we have a two-by-three-foot wall with ends and a corner. How about a larger wall? It is more of the same with a few exceptions. First, going much over three feet in height, you should increase the base or overall width by some eight inches for each added foot of height. The Height/ Width Chart gives an easy way to figure your width. For example, a seven-foot-high wall needs a four-foot eight-inch wide base.

Figuring Width

39

For a ten-foot height, the base should be seven feet wide. Width can remain at the base dimension right to the top or it can taper (to a point if you like).

Need for Footings The higher the wall, the deeper should be your footings. For any wall much over waist high (a major undertaking that presupposes an ample supply of stones), you should go down to below frost level or two feet, whichever is deeper. This is a lot of work lost to the eye forever, but it guarantees a good wall.

Then too, when you build above waist-high you get into leverage problems. The

Height-Width Chart. Vertical scale at left indicates wall heights (including footings). Read at right the proper width for the height planned.

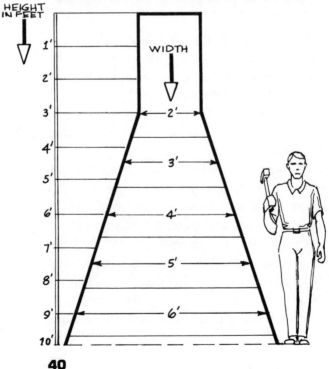

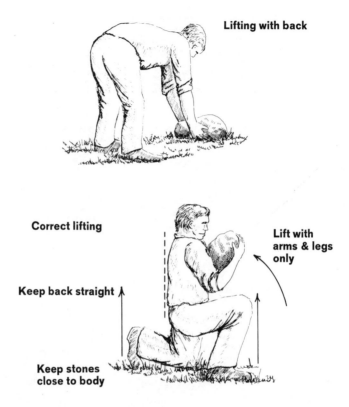

Incorrect lifting

Lifting with back

Correct lifting

Keep back straight

Lift with arms & legs only

Keep stones close to body

How to Lift

smaller stones can be manhandled, perhaps to mid-chest height. Just keep your
back straight and keep stones close to your
body. Lift with your legs and arms, not
with your back—at least not if yours is as
easily sprung as mine. Just one mishandled
stone and there's a sharp *ping* along the left
side of my hindquarters, and even turning
over in bed is agony for a week. Of course,
you can rupture yourself by lifting too
much the wrong way and that can mean
surgery.

41

1. Dig around rock. Use lengths of 2' x 8's (or similar. Block against side of hole to distribute load. Pry up one side. Drop stones under. Two people are better (naturally).

2. Pry up alternate sides.

3. Both pry.

4. Nudge rock on one plank and up with "muscle power." Can also use chain and winch or vehicle.

Chain attachment pictured at left will keep towed end from digging in. Attachment below will flip big stone over onto planking.

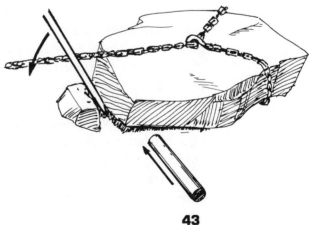

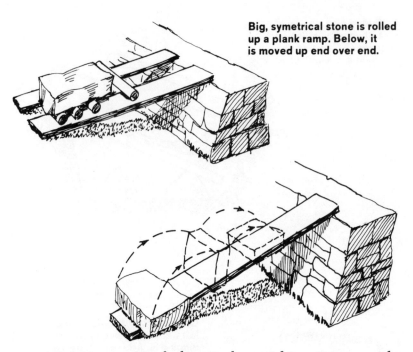

Big, symetrical stone is rolled up a plank ramp. Below, it is moved up end over end.

Mind, Not Muscle Much better than risking injury with extra-heavy stones or high wall work is to use the tools and rules of simple mechanics to do the work for you. The illustrations show how best to get a heavy rock out of the ground, and how to move one further along, by laying a road of planks and horsing the rock onto rollers.

Use the big crowbar to lever first one side, then another side up to get the first plank and then the rollers under. A ramp of planks, well supported with rock or stout posts, can serve to get the bigger stones up onto the wall. Just be sure that you have measured the available space and have done any needed trimming *before* moving the stone onto the wall. A dozen trips with the yardstick beat one round trip with a big stone because it failed to fit the first time.

44

Using Not-So-Good Stone

To this point, we've assumed you will be working mostly with good flat and square-ended stones. I hope you are so fortunate. But what if you find you've about half "good" and half odd-shaped, rounded or small stones—rubble? For example, good stone from an old barn foundation, plus a corn field-full of cobbles and pebbles. *Use Rubble?*

You can build a rubble-filled wall contained on sides and top by solidly-built "good" stone construction. Since this wall is as much a place to get rid of fieldstones as a boundary, you can make it as wide as necessary. Layout and footings are no different. But you want to build two narrow shells of good stone along each side of the outline.

Lay a course on each side, then fit in your rubble to the top. Make sure each cobble fits snugly on the rocks below and that the outer layers are tight against the outer shells. Tie this type wall a little better than an all-good-stone wall. A tiestone every two yards is not too much, alternating sides. *Fit It In*

45

A supply of fairly good stones is put together so as to invite problems. Note the excessive run to left of steps.

A completed wall. Note the use of the "one-over-two, two-over-one" principle.

46

On the next course reverse the pattern—that is, in course #1 you tie in from right, then left. Then, beginning to build from the same end, in course #2 you start on the left side, then tie in from the right, left, right, and so on.

Save your squarest rocks for ends and corners and use the flattest for the top. Try to make the top course all good flats, and, as always, the bigger the better. With rounded rubble under them the top stones are particularly apt to be dislodged.

Save Best Rocks

For this reason a good many rubble-filled walls are built with the top courses having two parallel, distinct sides, each dipping slightly down and in toward the center. So if a top stone is jiggled it can't slip off the wall.

This principle of down-and-in slant is carried even further in all not-so-good-stone construction. If you are so unfavored as to have almost all rounded fieldstones, you still can build a respectable wall.

It will not be as stable as if you had better stones, and lacking maintenance it can slowly turn back into a stone pile. Without footings and with sloppy construction and no care, a round stone and rubble wall can fall apart in as little as a hundred years. So dig deep, build with care, and leave funds in your will for perpetual wall care, so yours will keep standing.

Dig Deep

Essentials of construction are the same with all walls. The layout, footings, height/width relationship and tying-in hold for all stone walls. You will want to save all the flats for top stone, any with square corners for ends and corners, long ones for tying in. The better not-so-good stones (ones with one or two more-or-less flat sides) go in the outer shells, much as with the half-and-half wall discussed above. However, laying courses takes a new approach.

In or Out?   And herewith surfaces the Great Stone Wall Controversy: whether you chink in or chink out. That is: do you build up the core first, and put the outer shells on with any wedges or rubble plugs chinked *in*, toward the wall? Or do you build the shell, fill in the core and have wedges and plugs chinked *out*? Or maybe it's the other way around. Anyway I chink both ways. Half the best walls I know are chinked in, the

Shimming or chinking the outside of the wall.

48

other half chinked out. You do as you see fit.

Footings in this wall are laid in the pre- ferred slight "V" shape. Here it's more important than in other walls. The "V" is retained through all courses and the wall should have a slight inward slope to each face, even if it is low. The idea is to give gravity a bit more of an advantage so as to offset the tendency of a rounded stone to roll.

When good stones are few, fill center core with rubble.

The two flatest facets of each rock should be at top and bottom. *Most important,* be sure that the top facet of each stone has a slight downward slant on which the next course can rest.

49

Round stone and rubble walls

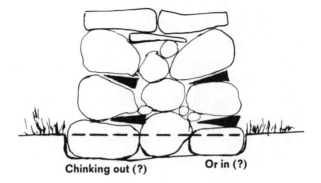

Chinking out (?)   **Or in (?)**

A sound wall can be made largely with rounded stones. The slant to center adds to wall stability.

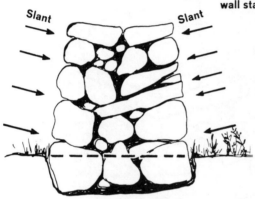

Slant   Slant

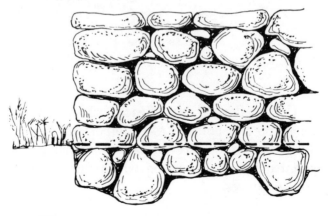

In building a course, I like to pile a line or two of the oddest-shaped rocks down the center of the wall, adding wedges and splinters to make sure they won't move. Then the outer face stones go on, each requiring a lot of juggling so that it rests firmly against at least five other rocks—both rocks to each side, two other face rocks in the course below and at least one big core rock inside.

It is good to fill as much of the wall as you can with small stones. Wedge-shaped stones should be saved to shim up face stones or fill openings in the face. It is best to chink such wedges "out"—that is, have the widest end of the wedge snugly in the core, the thin edge pointing out. Hammer a wedge into the face from the outside and it is sure to work out in time, unless it is aimed nearly straight down.

Small Rocks

Suffice it to say that round stone and rubble wall building takes a sort of sixth sense to tell which stone goes where and in what position. Only experience can teach that. I've been at it for some time now and still have to handle most rocks a couple of times before I find the right fit. But I can show you a few old-timers who can lay up a good wall from a stone pile they've never seen before almost as fast as they can pick up or lever out the rock.

Experience Helps

Drainage

Finishing Wall Once your wall is up, shovel the spoil—the sod and soil removed for footings—up against the wall's base. It will settle and provide a slope for rain water to run out on. Round stone and rubble walls in particular need special attention to drainage.

If you suspect that heavy fall rains or spring run-off when the snow melts will turn into a periodic stream cutting across your wall's path, provide drainage. A strong water flow can undercut a wall, but a ditch cut into the earth upslope of the wall will help guide the water where you want it.

What About Water? You may want to build a waterway into the wall. Just build in a hole running through the wall in the first above-ground course. Or you may put in a culvert of metal or ceramic pipe. Most elaborate would be to dig ditches and have them converge at the culvert.

Any amount of water flowing through a wall in time will wash and undermine it—hence the use of a culvert pipe is recommended. Seal around the pipe's uphill end so that the water doesn't work around it.

Wall built across a slope should be provided with a drainage ditch on the uphill side. Install culvert through the wall where needed.

Crossing a Stream

Unless you have a strong hankering to try your hand at bridge engineering, better leave off your wall at either side of a moving stream. If the gap needs to be closed to fence cattle, stretch several strands of barbed wire taut between posts set down firmly at the wall ends. Anchor the lower strand about a foot above the stream's normal surface, tight but capable of releasing if snagged by ice or debris during times of high water.

Special Walls

In some parts of the country there are abundant outcroppings, ledges or quarries of rock such as mica schist that splits easily (by man or nature) into beautiful slabs or into even-sized chunks or posts.

Rock Posts   The easiest wall in the world is made from rock posts set up vertically with their bottom couple of feet buried in the ground. Place the posts a bit closer than the width of what you want to fence in or out.

With a good collection of long, flat stones and chunks you can make a *lattice fence.* Alternate rows of squarish blocks and thin slabs, placing a support block wherever the slabs in the course above happen to meet. The height and length of openings in the block courses should be gauged by the purpose of the fence. Courses one foot high will keep in grown sheep. High walls will contain cattle or horses. No stone wall on earth less than ten feet high can keep in goats (or small children), or keep deer out.

Retaining walls to hold earth banks, to

Retaining Wall (mortared)

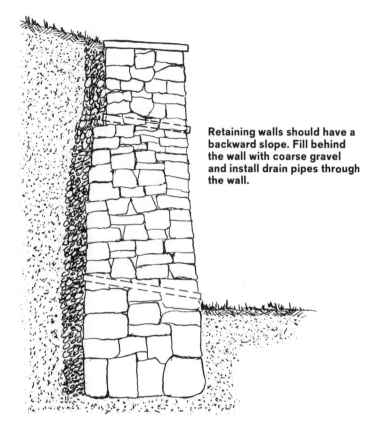

Retaining walls should have a backward slope. Fill behind the wall with coarse gravel and install drain pipes through the wall.

face earth dams or landscape a garden, are built just as any other stone wall, but the outer face must have a good backward slope to it. "Plenty of batter," as the old-time masons put it—the more angle the better; the higher the wall, the more angle needed.

Backward Slope

The inner face that meets the earth can be perpendicular or even slope backwards— possible only if you are building the wall against an existing earth bank, or fill behind the wall as you build.

Make sure the wall is well drained. Water can be trapped behind a retaining wall and bring it down. Best is to provide openings or culverts in the wall's base and pack a good foot of rubble in between the wall and the earth bank. Water then will be able to flow freely behind the wall and it won't build up pressure on any one spot.

Low retaining walls to raise or contain garden beds can be built to a height of four feet or so using only one or two runs of stone per course. Just set it into the soil at a good angle—45 degrees. Stones should overlap the course below for at least half their width. Pack good soil in between the stones and plant rock garden flowers, mosses and ferns in the wall.

Wall Furniture

With walls of any substantial height you will want to provide ways to get over, under or through. Simplest is to build a gap into the wall. For wide openings, you'll end up one wall and begin a new one. For smaller gaps you can continue the footing and perhaps the first course all along. But wherever you want the gap, build two good ends as far apart as you wish the gap.

Gates and Stiles

An old trick in cattle or horse fences is to build a narrow man-sized gap into the wall, but put it in at an angle, say an 18-inch-wide gap put into a three-foot-wide wall at a 45-degree angle. An animal looking square at the opening on its side of the fence won't see the daylight at the other side of the gap and won't be tempted to try to squeeze through or work at the stones to dislodge them.

Fool the Animals

Another way over a stock fence is a *stile*. Here you build extra long, narrow and flat-topped rocks into the wall so they stick out for a foot or more. Arrange them in ascend-

Built-in Steps

ing order at an easy climbing angle and distance apart. Your average stair rises about eight inches and moves eight inches forward per step. You can increase this to a foot in both directions. Be sure to plan this stile—both up and down flight—before you begin building.

The top rocks holding on the highest stair-stones must be extra large to hold these stones in (and hold you when you climb the stile). The stair-stones must be especially well lodged in the wall. You may even have to cheat a bit and mortar them in. More on mortar later on.

Open Stairs

No farm animal but a goat can climb a step stile, but if you have no need to contain

Stile steps set into the wall on both sides.

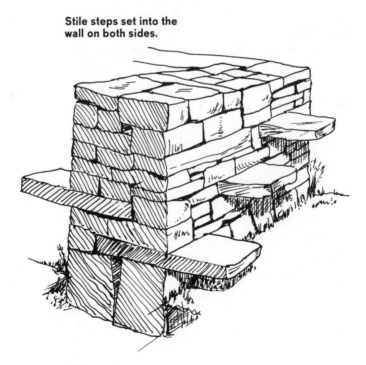

stock you can build an open stair into the wall. If the wall is wide enough, you can build both flights at the same spot in the wall. More likely, you will have to build one flight into the wall at one place, the other a bit further on.

Flights of stairs are perhaps the wall builder's greatest challenge—and monument.

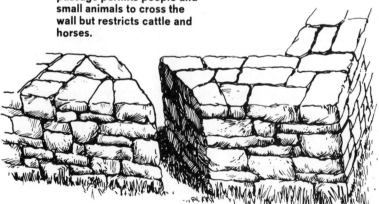

A diagonally built narrow passage permits people and small animals to cross the wall but restricts cattle and horses.

Low, easily laid retaining walls of small stones create a lovely backyard rock garden and a convenient place to sit and shuck sweet corn.

There's nothing special to building stairs. You just leave out stones farther and farther in as each course goes on. Treat the stair walls as you would an end, and use especially square-cornered stones for the steps themselves. Each step should go up and in something less than a foot.

Gate and Latch   You can put a gate and latch in the pasture-side stair and it will hold most live-stock. Gates, naturally, can go into any gap in your wall. Traditionally people gates in New England walls are made of wrought iron. Farm gates are wood, as a horse-team-wide gate of iron would weigh too much and the upper hinges would rupture too

often. You can make or buy a gate of whatever material and design suits your fancy.
The problem is fastening gate and latch to
the wall.

You can bury stout gate posts of creosote- *Use Gate Posts*
soaked wood in the ground at each side of
the gap in your fence, then attach the gate
hardware with wood screws. If the gate is
more than man-high, you can frame it in
with wood, brace side posts with a lintel at
the top and threshold at the bottom and
hang a regular door.

More in keeping with the craft, I think, *Attaching Bolts*
is to put your gate hangers right into the
wall. One way is to plant heavy bolts high
and low into the facing ends. Have the
thread protruding three inches out of the
wall. Then you can bolt on two-by-fours
with holes drilled at the appropriate places.

Bolts don't have to be precisely in line,
as you can make the gate post plumb by
adjusting spacing of the holes and putting
washers on one or another bolt. Then just
attach your gate hardware. You may have
to cement the bolts in.

You can set or cement in eyebolts also, *Use Cement*
two on the hinge side, one on the latch side.

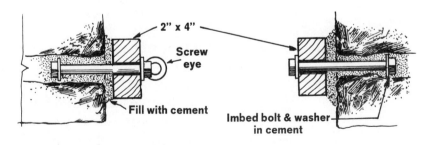

2" x 4"

Screw
eye

Fill with cement

Imbed bolt & washer
in cement

Hang the gate from the eyes with a pair of open-ended hinge pins. Or put four more eyebolts into the gate's spine—one just above, the other just below each eyebolt in the wall. Then put spikes or bolts through each set of eyebolts.

Another pair of eyebolts in the latch side can be set to line up with the single one in

Sturdy gate anchors may be set between wall stones using mortar.

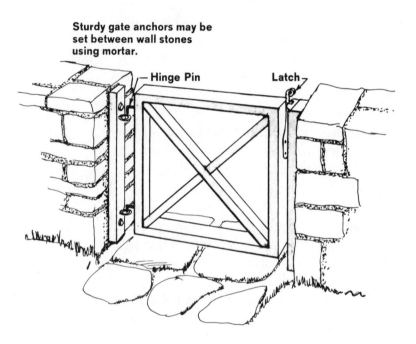

— Hinge Pin   Latch —

the wall end at that side of the gap. Latch with a bolt, spike or length of wood.

Drilling Holes   The most workmanlike method is to set gate hangers, eyebolts usually, into holes drilled into the stone itself. It is best and easiest to do your drilling before the ends are laid up; trying to drill in horizontally is tiring, especially low down. To drill you need your narrow stone chisel, a heavy-headed hand sledge or mason's hammer and a dust scoop. The scoop is to get rock dust out of the hole. A sufficiently long screwdriver works; you just keep scraping dust up and out in small dribbles. Better is to make a real scoop. (You can't buy one anywhere that I know of.) It is simply a length of soft iron, more or less screwdriver-shaped, but with the bit end flattened and bent up into a slight curve. This gadget can get out twice as much dust per lick as a plain iron. Hammer your own from soft steel rod from any hardware store.

The Dust Scoop

To drill, select your spot, put on the chisel and begin hammering. After each blow twist the chisel around a few degrees. You'll find that the chisel and hammer will spring up after each blow.

In time you can develop almost an effortless technique; the chisel will pop up away from the rock and you can give it a little twist with your fingers. The combined recoil of the tools from each blow will send the hammer half way back up and all you

8"

Dust scoop

need to do is give it a little boost each time and guide the process. Drill in at least two inches. Keep the dust out of the hole or you are wasting time and effort, but not getting any deeper into the stone. If you think of it, try to drill at a slight angle so the hole will head slightly down. And cut a slot in the hole as deep as you can. Then the hardware can't ever pull out or twist in the hole.

For Keeps   You can put in your hingepins or eyebolts with mortar, epoxy cement, can even wedge them in with shims of hardwood and they will stay put for a summer with a light gate. The best way is to wedge them in with saltpeter-water mortar or better with lead, just as plumbers lead-in junctions in iron soil pipe. Get a good punch that is long enough and small enough around that it can get between the pin or bolt all the way to the bottom of the hole. Then punch in strips of ordinary lead. If you tamp it in well all around, in effect filling what part of the hole the bolt isn't occupying, the bolt will stay till it rusts through. If you like you can use one of those little portable propane

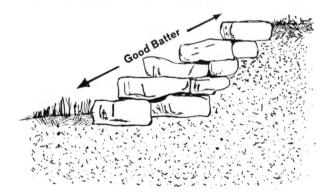

Good Batter

torches to melt the lead. Aim the flame at the bolt; it will heat and melt the lead around it.

Be sure to wear your gloves, goggles and a heavy jacket doing this. All rock contains some water, even if only tiny dribbles in small surface fissures. It can turn to steam when heated, can expand and blow a solid granite block wide open or blow out a chip. The risk of injury is slight with proper clothing and goggles. Still, I will stay with hand-tamping the lead and advise you do the same. *Careful*

Quarrying

The best tool for quarrying stone is dynamite placed by a licensed expert, and most is done in huge quarries. If you can't pick up your stones, but before setting out to hand-quarry, think again about buying a couple of truckloads of stone. It is normally sold by the cubic yard, much of the cost depending on delivery distance. To find the number of yards you need, multiply the *Figuring Needs* length of your projected wall by its width. Then multiply the resulting figure by the wall's height from bottom of the footing to the top. If you've done your arithmetic in feet, you have the number of cubic feet in the wall. To find the number of cubic yards, divide by 27. If you'd like to sound old-timey, you can order rock by the *perch*, based on an old surveyor's measure. A perch is a stone wall $16^1/_2$ feet long, 1 foot wide and $1^1/_2$ feet high or $24^3/_4$ cubic feet, something less than a cubic yard.

If you'd like to, you can go ahead and take a crack at hand quarrying. You will need several really big sledges, at least one with the mason's pointed head. Several single-bladed ax-sized wedges, some mason's points, a farm tractor and trailer (or a truck with power winch and the ability to get to your rock and back) plus fifty feet or so of logging chain, an assortment of stout planks, rollers, your safety goggles, boots and gloves. And, of course, a big lunch (and plenty of drinking water in hot weather). *Tools Needed*

You may also try to find an old stone boat, an oversized skidding sled you can just roll rocks on and off of, that is good for hauling rock for short distances over fairly level, grassy land. *Moving Stones*

Next, find an outcropping of rock, or dig down to a buried ledge that perhaps reveals itself by poking just a tip through the soil.

If a tap with a hammer leaves only the barest nick and/or the hammer makes sparks, you likely have igneous rock—especially if the stone under the weathering or lichen shows small hard, shiny crystals of several dull colors. These rocks do have a grain and can be split with difficulty—*considerable* difficulty.

If this art appeals to you, find a quarry boss, sculptor or gravestone carver to teach you how to make out the grain. And be prepared to spend hours drilling holes and working in your splitting wedges. Most old New England farm houses rest on split granite sills that have the hand-drilled holes showing. But those were hardier times. *Slow Work*

If you find a softer rock, look it over for lines, streaks or patterns in the exposed

67

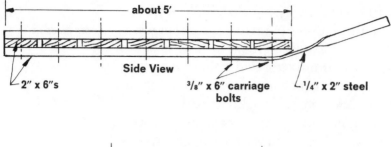

about 5'

Side View

2" x 6"s

³/₈" x 6" carriage bolts

¹/₄" x 2" steel

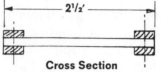

2¹/₂'

Cross Section

Plan and materials for making stone boat, shown below.

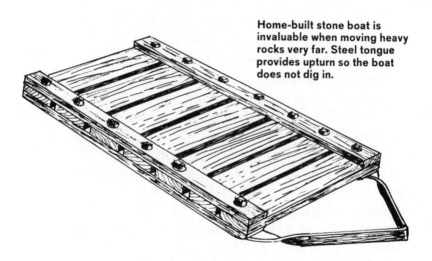

Home-built stone boat is invaluable when moving heavy rocks very far. Steel tongue provides upturn so the boat does not dig in.

face that suggest a grain. Sometimes chip-ping off a piece will produce a flake; likely the grain lies along the plane on the rock where the chip came off. Other rocks will have definite cracks or streaks indicating grain or the interface between two splittable strata.

Seldom will a sedimentary or igneous rock have grain that runs absolutely horizontal or vertical. It will go back into the earth at odd angles. Two terms describing these angles (that you don't need to know but which will make you sound like a real miner) are "dip" and "strike." *Dip* is the angle with the horizontal that a stratum makes. The strata are more or less flat planes.

"Dip" and "Strike"

If you drew a line along where the strata meet a horizontal plane, the direction on the compass that the line heads off in is the *strike.* Tell an old-time stone cutter the dip and strike where a particularly good splitting stone dives into the mountain and he'll be able to find where it comes out on the other side.

Once you have these angles, you know what you will be splitting off of what and where to cut in. Start with the big mason's hammer, hitting hard in a line along the grain with the pointed head. Hammer a good dent out every six inches and keep going up and down the line. In time a crack should begin to open. Insert points, then wedges in the crack and keep pounding. When the crack is large enough, insert the crowbar and, with a little back muscle, a slab should break out.

Keep Pounding

Use your vehicle and the logging chain

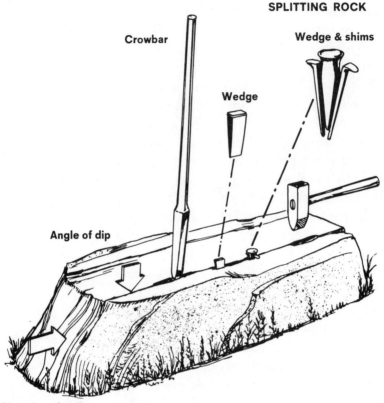

Crowbar

Wedge & shims

Wedge

Angle of dip

Direction of strike

Moving the
Big Ones

to move really large chunks—you will probably have to lever the rock up, support it with smaller rocks to get the chain around and under. Then pull it out by brute force if your vehicle is up to it. Otherwise, lever it up, slip in rollers and move it that way. If the ground is uneven, you may have to make a track for the rollers with your planks.

Some slabs will break into building-size pieces by being hit repeatedly with a sledge. You may have to tip a slab up against the hillside or put one edge up on rocks or a

70

log so it is unsupported in the midsection and more apt to break. Other rocks must be scored along the entire splitting line and supported on each side of the desired break, the split made by more sledge work.

Besides being *very* hard work, quarrying can be dangerous. Never try to lever up a large slab alone. It could slip off supporting stones or logs, pinning a foot, hand or arm.

Watch Out

When working the face of any high formation, keep looking up and listen for falling rocks. Each stone you remove changes the natural stress deep within the formation. It's no exaggeration to say that one good and successful tug with a crowbar can bring the whole face of the cliff down on you. And the easier the rock is to remove, the more likely the chance of trouble.

Walls of easily pried-off "rotten" stone, hills with a lot of loose chips in piles at the bottom, and formations where the dip is nearly vertical and the strike runs more or less even with the cliff face are good places *not* to quarry.

Avoid Rotten Stone

Mortar

If you absolutely must, you can put mortar in a stone wall. Buy several sacks of ready-mixed dry mortar (not sand mix or concrete mix) and add water per the directions till you have a good workable mortar. Mix only the amount you are sure you can use in a quarter or half hour. Take mortar from the edges of the barrow or mixing bucket and keep it stirred up to prevent premature setting.

Problems with Mortar All open spaces in a wall can be filled with mortar—and in New England winters the mortar will crumble and fall apart before a properly laid dry stone wall shows the first sign of age. Dry stone walls are laid to give and take with changes in temperature. A mortar wall is rigid, and since it can't absorb stress—move with the season, so to say—it *has* to break up.

You can throw mortar into the interior of a rubble-filled wall and build the outer face and top dry. If you are low on rocks, use stone only for the top and facings and make

72

the interior of clay, gravel or stoney soil mixed with water and straw. Pack it well and fill in, or "point," the cracks between outer stones with mortar. It is best to cement the entire top of such a wall. That is, set the top rocks into a solid bed or mortar.

In all trowel work, make sure the rocks *Wet the Rocks* are wet before applying mortar. Make sure all pointing has a down and out slope to it,

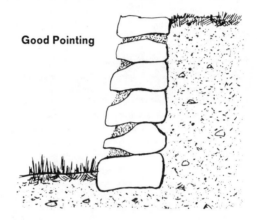

Good Pointing

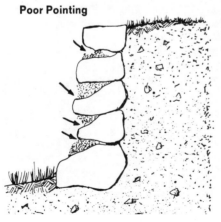

Poor Pointing

Good pointing in a mortared retaining wall (above) sheds water. Pointing at right invites moisture, freezing and cracking.

Chunky and irregular stones become a handsome mortared wall screen for a secluded city home.

Mostly irregular stones here are laid up to make an attractive and durable retaining wall. A top course of heavier stones would have been better.

74

Wall built with big cobble stones cemented together.

and cut water channels in large expanses of mortar. The objective is to guide rain water out of the wall, particularly to keep it from getting between rock and mortar. If you leave a pit or lip or fail to mate rock and mortar well all around, water will stand and leach out your pointing. If it freezes, it will depoint the whole wall in a few winters' time. All it needs is one toehold to get started.

Mortared retaining walls require special drainage planning. The interior rubble layer mentioned earlier is more important, and drains—ceramic pipes from rubble layer out to the face of the wall—must be laid along the wall bottom, and in a row ten feet or so apart half way up if the wall is more than ten or twelve feet high. Place other rows for every five or six feet of additional height.

75

Maintenance

A well-built wall deserves the small amount of annual care needed to keep it standing. Each spring I patrol our walls that range in age from 200 to 2 years of age, replacing stones knocked off by weather or visitors. You get to know your walls this way, get to know good and bad points of construction too, as the same faults crop up in the bad sections year after year.

Clear Out Growth It is not a good idea to let vines or brush get well rooted in a wall. The roots can slowly force stones apart. I cut back the grape vines, poison ivy and assorted brush plants that invade the walls each fall. I also try to do away with any woodchucks that take up residence in a rubble-filled wall. A live trap baited with a few cabbage leaves will usually capture the intruder in a day's time.

If you have mortared walls, check for damage each spring. Don't ever put back loose mortar. Dig it all out and put in new. Check any drainage pipes annually and remove the birds' nests, leaves and sticks that will accumulate and clog them up.

And finally, you may want to give in to your natural pride at building so substantial a structure as a stone wall, and sign it. I do.

Havahart trap

I don't go so far as to set up a sign that reads "This Wall was Built by John Vivian." The small chisel will carve your initials and the year quite nicely. Just chip out letters and numbers in v-shaped grooves. Sign the biggest rock in your best corner, end or stair —in full public view or off in a private cranny depending on your nature. Good wall building to you.

Some Other Uses of Stone

By Ralph Scott

Stone walls, wildlife, and the natural landscape go together. Unlike other forms of
"fencing," stone walls lend a sense of harmony to the land. As such structures "age,"
lichens and mosses decorate the stones in a
mosaic of subtle greens and grays. Vines
climb over the wall, anchoring their tendrils
in the multitude of interstices. Trees and
shrubs invade the wall's foundation, adding
vertical dimension to the horizontal tra-
verse. As the wall ages with its succession
of plant growth, many birds and mammals
take up residence amid the rocks and niches
of the invading vegetation. This section of
Building Stone Walls will emphasize some
additional uses of stone as a natural building material for helping the land and its
wildlife.

Stone Dams   As we shall see, stone walls are not only

78

effective conservation structures for the land, but for water as well, particularly streams. That's right, you can build stone walls (or modifications of them) in streams!

Simple stone structures—dams and deflectors—carefully placed in a waterway will increase its productivity of fish, particularly such highly prized game species as trout and small-mouthed bass.

Fish, like birds and mammals on land, require food, shelter, and living space to thrive. These aquatic creatures also require rather specific stream bottom or substrate types for spawning.

By building stone deflectors and dams in flowing waters, all these needs are increased through the intentional modification of the waterway's environment.

Deflectors

What these stone structures actually do is break up the stream into a series of fast-flowing riffles, alternating with slow-moving pools and deep channels. The current's velocity is altered and it is essentially bounced back and forth from bank to bank, causing the bottom to deepen in some places and depositing sand and gravel bars (or shallows) in others.

Provide Pools

Before starting a stream improvement project using stone, examine the accompanying illustration. This drawing shows a stretch of stream and the various types of dams and deflectors one can make, plus the series of small fish habitats each structure is likely to produce.

There is no formula as to how many of

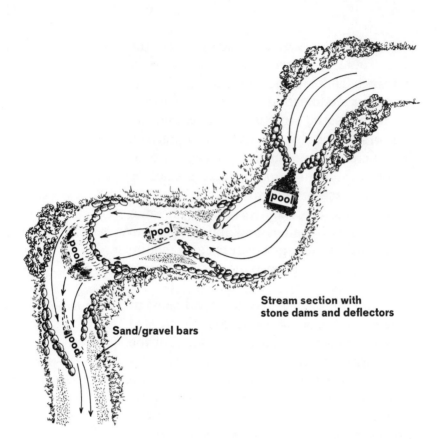

Stream section with stone dams and deflectors

Sand/gravel bars

these structures should be placed in a waterway. This, of course, depends upon the existing conditions of the stream—number of meanders, availability of stone, type of bottom, fall and velocity of current and so on.

Protect the Banks   The only important precaution to take is to observe carefully how you have modified the current's flow, especially in bouncing it back and forth from bank to bank. Make sure that each bank receiving the current is adequately protected from erosion by another stone structure or vegetation, such as a thicket of willows.

The actual building of these stream struc- *Tie Stones*
tures is basically the same process as con- *Needed*
structing a free-formed stone wall. It is especially important to place "tie stones" properly so the dam or deflector can withstand
the force of the waterway's current.

When beginning the construction of a
stream structure, the condition of the bottom is not critical. It is often possible to
build upon boulders that naturally form
small dams, deflectors, or riffles, placing and
tying in additional stones to the desired
height, which, in most streams, varies from
one to two feet (sometimes three feet).

If you prepare the stream bottom with a *Make It Stable*
foundation of small stones and gravel, however, the dam or deflector will possess greater stability. Note also that these structures
are built in a somewhat triangular-shape
when viewed in cross section (next page),
with the "isosceles" of the angle pointing
downstream.

This shape gives the structure added stability, and there is a boulder riffle created
on the downstream side where a multitude
of aquatic insects and other invertebrates
can attach. These are for the basic food for
the sport fish you wish to produce.

Another feature of these mini-stone walls
is that they are bowed. That is, when laying
the stones be sure to make the structure form
a slight bow toward the downstream side.
This will also reduce the chances of wash-
out and lessen the force of the current
against the dam or deflector. However, anyone building these fish habitat structures

81

should be prepared to perform some annual maintenance and to replace stones. Heavy spring rains and runoff, resulting in annual high waters, will almost always move some of the rocks.

Watch for Change   Within a year, you should observe some changes in your improved stream. New sand and gravel bars will be formed. Quiet pools will begin to deepen below small dams and double deflectors. As the structures age, just as stone walls on land do, algae will form on the rocks. Fish food, in the form of aquatic insects and other invertebrates, will hide amid the tangles of algae. Then, as the population of the food organisms increases, so will the fish population.

Then there will be the day when, through the clear, well-oxygenated water, you will see trout or bass feeding and resting in the pools below the structures. You will be able to cast a fly into the cool waters, wait for a strike, and land your much worked-for

Cross-section of a stream dam/deflector showing triangular shape

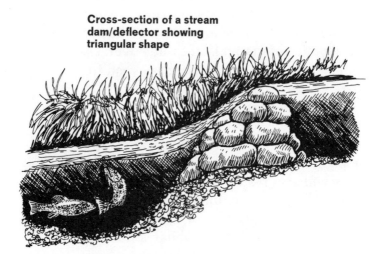

prize. You will learn where the fish feed, where they hide, rest, and spawn—for you have improved the stream habitat; you have built stone walls in waterways.

The use of stone in aquatic environments is not limited to streams. You can use it for protecting the earthen dikes and dams of your small pond.

Invading vegetation, particularly willows, and muskrats are a plague to earthen dikes and dams. Cattails, bulrushes, and willows will weaken an earthen structure with their root systems. This is particularly true of willow trees if left to grow. Muskrats will burrow into earthen dams several feet. These homesites are eight to ten inches in diameter and are serious threats to the stability of an earthen structure.

Stone, used as *rip-rap*, will help alleviate nature's in-roads to dams and dikes. Relatively flat stones are preferred over round boulders, and these should be placed as close together as possible to reduce spaces in which vegetation can gain a foothold.

Two Layers   When rip-rapping the face of an earthen structure, two layers of stone are recommended—a base layer and another layer on top which overlaps the spaces of the first layer.

To protect the waterside of an earthen structure, one must know the maximum and minimum water levels of his pond. Thus, the rip-rap should extend one to two feet *above the high water level;* and two to three feet *below the low water level.* This will not only protect the structure from burrowing muskrats and plant invasion, but from wave action, which can erode the face of an earthen dike or dam.

Top view of rip-rap (2 layers), showing overlapping of stones to reduce spaces where plants can take root.

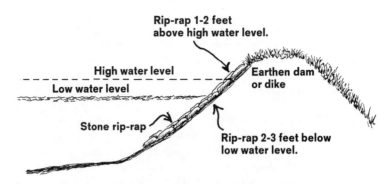

Rip-rap 1-2 feet above high water level.

High water level

Low water level

Stone rip-rap

Earthen dam or dike

Rip-rap 2-3 feet below low water level.

Dam-cross-section showing upper and lower limits of stone rip-rap.

If your property has some large erosion gullies due to some past land abuse, stone walls or dams can be used to check the continuing erosion process and greatly reduce further soil losses. These dams are properly called "check dams," and have been in use for many years by soil conservationists and technicians.

Stone Dams for Erosion Control

Such erosion control structures are relatively easy to build, and they are not much different in construction than a stone wall. The same basic building guidelines for a wall apply to check dams, the placement of tie stones and footings being equally important for stability and support.

There is no set number of check dams you should build in a gully. The number and spacing between structures will depend upon the size, length, and grade or percent slope of the erosion feature being brought under control.

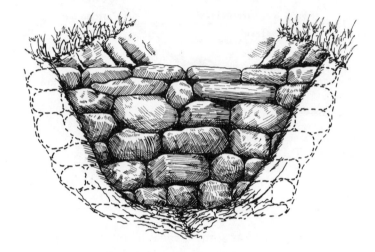

"Anchoring" check dam into banks to prevent washing around structure and widening of gully. Dig check dam in a foot or two, cover with soil and pack.

Start at Bottom Build your first check dam at the base or bottom of the gully. Place the footing stones as recommended in the section on wall building, and dig into each bank of the gully to a depth of a foot or two. Pile the stones as previously recommended to a height at least level with the top of the banks. Each layer of rock or "run" should go into the sides of the banks you excavated, and be covered and firmly packed with soil. This will serve as an anchor for preventing runoff water from washing around the structure, causing the gully to widen and further erode.

Check the Grade The placement of the next, and succeeding, check dam will depend upon the grade of the gully. Use a carpenter's level, placed on top of the completed dam, and held level so the bubble is centered. Sight across the top to determine the position where the next check dam will be placed. (The placement of wire brads or pins on each end of the carpenter's level will facilitate easier and more accurate sighting.) Have a helper move up to the next location, and set a stake at the point where your sighting is positioned. Thus, the base of the next check dam will be level with the top of the structure just completed.

For each check dam, repeat the building and leveling procedure up the gully to its beginning or head. As you progress, the stone structures will become progressively closer together, as well as requiring a shorter "run" or length to dam the gully's width. *Closer Together*

After the series of small dams are in place, runoff waters will be drastically reduced in their velocity and erosional power. Pooling will occur behind each dam, where the silt

Levelling with a carpenter's level for determining placement of check dams in gully.

Stake driven at base point of next check dam

Line of sight

Carpenter's level with bubble centered

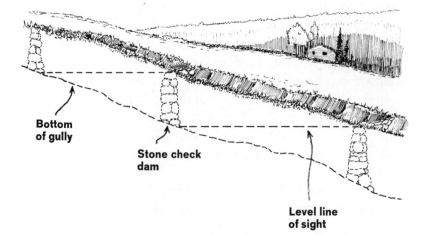

Bottom
of gully

Stone check
dam

Level line
of sight

Cross-section of check dam placement "on level"

loads will deposit and gradually accumulate as the water slowly drains through the spaces (and sometimes over the dams in a real "gully-washer") in the stone check dams.

Vegetation Will Help   In time, vegetation will gain a firm foothold (or you can plant trees and shrubs to speed up the recovery process) in the accumulated soil and further slow down storm runoff, reducing erosion. As the vegetation develops and matures, the gully will become a brushy haven for songbirds and wildlife, as well as a healed scar upon the land.

A Project for Left-over Stone: Build a Bird Bath   After you have completed a stone wall or any other stone-building project, you may have some large boulders left over. These may be either surplus or those that just

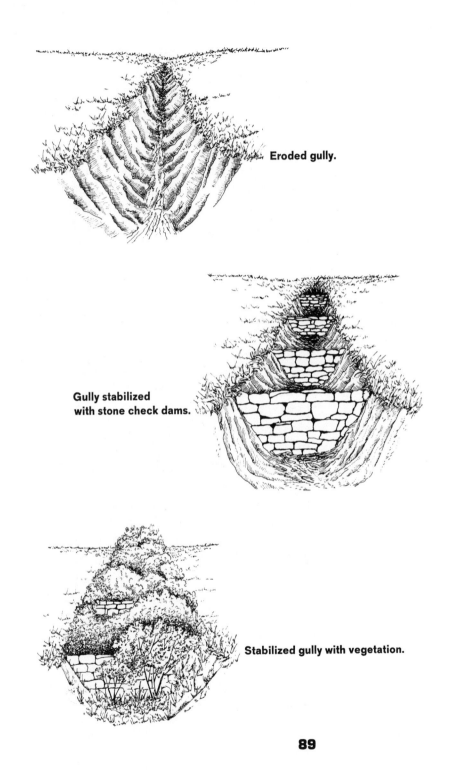

Eroded gully.

Gully stabilized
with stone check dams.

Stabilized gully with vegetation.

89

didn't seem to fit. Well, put them to use in your yard or garden. Build a stone bird bath.

Attracting songbirds to one's homesite is a common practice among gardeners. Not only do these avian friends enrich our daily lives, but they are an asset in controlling insects and other garden pests. Thus, many gardeners provide nesting boxes, food-bearing and cover-producing trees and shrubs, and supplemental feeders. But, one elemental need of birds—water—is often overlooked.

Expect Visitors By adding a stone bird bath to the garden habitat, you will encourage many species, some rarely observed, to be regular visitors to your homesite. And if the water you provide shows some action—a steady drip or trickle—it becomes even more attractive.

Begin your stone bird bath building project by assessing your garden's habitat. Because birds tend to approach a bathing area in steps, adjacent cover is important for them to hop shrub to shrub and branch to branch. However, the bird bath should not be placed so close to adjacent cover that the family cat or other predator can lie in ambush. Birds are least wary when they are bathing. Also, place the bird bath so it receives sunshine for at least half the day.

The height, size, and design of your bird bath will depend upon personal preference, plus how many stones are left over or available.

The accompanying illustration is an easily constructed stone bird bath requiring the least amount of accessory materials.

90

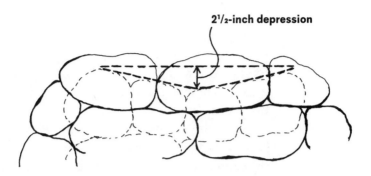

2¹/₂-inch depression

Cross-section of stone bird bath with cement pool.

The stones are piled like a *cairn,* leaving a small depression in the center where a cement bottom is formed for the pool. The placement of the stones is not critical, so long as they are balanced and can be supported so the whole structure does not topple over. However, the use of mortar with good pointing (see the section on "Mortar") will insure the bird bath's permanence.

A foundation of gravel and small stone is recommended for the bird bath to rest upon. This will also serve as drainage around the structure, which is particularly important if you plan to have a steady, dripping water source. Eventually, such a steady water supply will overflow the pool capacity and reach the ground. Also, certain birds, particularly grackles and jays, are notorious for their splashing. Such vigorous bathing often empties the shallow pool and the water flows over the sides. For this reason, a foot or two of gravel completely encircling the

Needs Foundation

stone bird bath is recommended. Otherwise, be prepared for mud.

In building the foundation/drainage pit, excavate the desired area to a depth of four to six inches and fill with crushed rock or gravel. Then begin to pile the stones for the bird bath.

Gradual Slope After the stones are piled to the desired height and design, make the cement pool in the center depression. This pool should slope gradually from the sides toward the center to a maximum depth of $2^1/_2$ inches. Birds prefer to bathe in shallow water, and most garden songbirds fear water depths greater than a few inches. To make the pool any deeper than recommended often defeats its intended purpose. Many birds, particularly the smaller, insect-eating species, will simply avoid your watering device. Also, be sure the bottom of the pool is rough (no need for smooth, expert masonry work here). This insures that the bathing birds can maintain firm footing.

Dripping Water If you plan to supply dripping water to add motion to the pool (which is recommended since the motion attracts a greater variety of species), then you may have to cement-in a three- to four-foot, bent steel rod. This should have a loop at the end overhanging the pool in which the water supply source is hung or attached. Of course, if your bird bath is placed where an overhanging branch is readily available to serve as a support, then a rod will not be necessary.

Simple Method There are two, inexpensive water-drip sources available. The cheapest and simplest is a two-gallon bucket with a very tiny hole punched in the bottom. This bucket hangs from a tree limb or steel rod support and is filled with water, which drips slowly into the pool.

Another method (see Page 94) utilizes several feet of 3/8-inch plastic tubing connected to a *reduction coupler*, which is attached to a garden hose connection and threaded onto an outside faucet. The length

Use Plastic Tubing

Bucket-drip water supply for stone bird bath.

Typical stone cairn-style bird bath with drip hose.

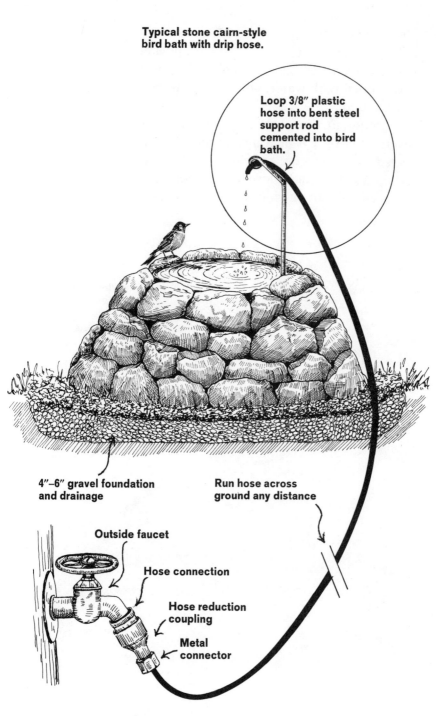

Loop 3/8" plastic hose into bent steel support rod cemented into bird bath.

4"–6" gravel foundation and drainage

Run hose across ground any distance

Outside faucet

Hose connection

Hose reduction coupling

Metal connector

of tubing required will, of course, depend upon the distance from the water source to the bird bath.

Run the tubing across the ground and loop it through the rod or over the limb. Turn on the faucet so a slow, but continuous, dripping of water will create the noise and motion you (and the birds) desire. At the time of this writing, such a rig could be put together for about $7.00 (based on 20 feet of plastic tubing). The hose reduction couplings are sometimes hard to find, but are generally available from a plumber's supply house.

Continuous Drip

The Story of a Wall

By Roger Griffith

A Wall—
A Monument This is the story of a wall and the two brothers who built it. The wall is a monumental one, three feet high and four and usually more feet wide. It runs about 600 feet along Route 30 in South Dorset, Vermont, and there's another section of 200 or more feet that is at right angles to the highway. Sections of it are more than 50 years old but it still has a neat look of newness.

The Burns brothers, William T. and the late James, built it over a period of 10 years, working on holidays and vacations. As someone told Bill one day recently, "That wall, it's quite a monument to your brother."

A Supply of
Stones To begin this story at the beginning means going back many years, back to the Ice Age when the glaciers were scouring up and down Vermont, grinding into the bed rock, and occasionally melting and spewing huge bulks of sand and rocks and debris down on the countryside. Just such a jumble of material was dumped into the valley

96

now shared by South Dorset families and the West Branch of the Battenkill.

Now jump ahead a few millennia. That brings you up to the time before the Revolutionary War when the Underhills settled in South Dorset and began to scratch a living out of that rocky soil. The scratching produced food for a family and its livestock—and rocks. Each year more and more rocks that the plows brought up had to be moved out of the way of the corn, the potatoes and the grain. Other families followed the Underhills on the land, and each did the same thing, plowed the level fields, then hauled that year's crop of stones to the side of the lot, using a stone boat, or snaking the big ones with a team of horses.

By the time the Burns brothers' grandfather bought the property, just after the Civil War, there was a triangular pile of rocks, waist high and stretching along the road, and another running near the farmhouse from the road to the brook down in back of the house. — **The Raw Material**

And that brings us up to the summer of 1921. Bill Burns is home from the University of Vermont where he's studying engineering. He's playing semi-pro baseball in North Adams and Bennington and coming home to South Dorset between games. He's a husky six-footer, capable of picking 300 pounds of rock off the ground without grunting. He's itching for more exercise than he's getting and wants that exercise to be productive. — **Ready to Work**

Bill knows about stonework. His father

97

has worked all his life in the marble quarries and marble mills that are almost within shouting distance of their home, and Bill himself, in a year off between high school and college, worked to raise money in the old Kent and Root quarry a half-mile or so down the road.

Drop a stone into place, he says.

So, in this summer of 1921, when his mother wishes out loud that she had more space for flowers in the yard south of the house, and a neat wall instead of that pile of stones and jungle of brush between the yard and the pasture, Bill gets the idea of building the wall.

Started in 1921

He didn't get any advice on how to go about it. "I just started building," moving first one rock, then another.

Bill has retired now, after a career in engineering that included being vice president of the Kings County Lighting Co. in Brooklyn. Sixteen years ago he and his wife Elinor moved back to South Dorset and into the house across the road from where he was born and where his sister still lives. The location across the road is good for Bill—there's a better view of the wall from there.

Good View of Wall

Even after these 50 and more years, the details of constructing that wall haven't dimmed in his mind.

"There was a big pile of rocks, so we didn't have to lug any in," he explained. "We took rocks from in front of us, built wall in back of us, leaving an open space in between for working."

He attracted attention. "Old quarrymen would stop and visit. 'Put one on two.' It was the only advice they would give me." It meant to avoid balancing one rock on top of another, for someday the forces of man or nature would upset that balance. Instead, make certain every rock rested on two or more other ones, for stability. Bill found himself almost singing that advice as he

Got Advice

99

lifted and placed the huge stones: "Put one on two. Put one on two."

Some weren't encouraging. "One fellow, who used to come here summers, stopped and watched me. He said he was working on a wall, too, but said he was using mortar to hold the stones together. 'That wall of yours won't last five years,' he told me."

How has that other wall stood up over the 50 years?

"It cracked," Bill reported, with a grin of satisfaction.

Bill learned many lessons the hard way, and now, after 50 years, can tell, by looking at the wall, which of his methods were best.

His Advice One piece of advice he gives is to "Put your big stones on the outside of the wall and at the bottom. They anchor themselves pretty well that way. Then put the smaller ones in the middle. It saves lifting those heavy stones and besides it's better for them. They don't move any more. If you put smaller ones or round ones at the bottom, then put a bigger one on top of them, sometimes those at the bottom will sink, with one sinking more than the other and throwing the wall out of line.

"If a stone has a flat side, I'll put that flat side down, too. Some builders will put the flat side out, to make the wall look better, but if you do that the stone is sticking into the wall and it will slide out in time."

Help From Jim Bill's younger brother Jim returned to South Dorset in 1930 after playing semi-pro ball in Florida and serving on a police force there. He joined in the wall-building.

"Most of the time I worked alone, and Jim worked alone, because we weren't here at the same time. When we were both here, he always wanted to play golf. That's when I learned that if we worked hard at the wall for two or three days, I could beat him at golf because he wasn't as used to the work on the wall as I was."

Bill and Jim didn't quite agree on wall-building techniques.

Didn't Agree

"Jim put more uniformly sized stones on the outside. His wall looked better than mine because he would hide the big stones in the middle, while I put the big ones on the outside."

Today Bill can walk the length of the wall and say, "Jim built that section, and I built along here."

Which method has best stood the test of time?

"I think the sections where I put the big ones on the outside, because I've seen that if the big stones are in the center, the others will slide away."

Another lesson learned painfully was not to set a rock down on the wall, but to drop it. "If you set them down, sooner or later you'll pinch your fingers. So drop them about like this,"—and he dropped a rock about six inches. "It will bed better too, that way."

Painful Lesson

Bill advises against shimming the bigger stones with smaller pieces. "At first the wall looks better and the stones look nice and square," he conceded. "But in a few years those shims will start falling out and then the larger stones will fall out. Once in a while I'll put a shim in, but on the inside of the wall where it is locked in place and can't fall out."

Won't Last Bill is watching a wall only recently built in his neighborhood. "It looks nice now, but in a few years those shims are going to start falling out," he predicted.

Laid By Eye Bill and Jim never used strings to guide their rock-laying. "We would put up two stakes, one for the inside of the wall and the other for the outside, put them up maybe 50 or 60 feet away. As you laid, you laid by eye, to the stake, and every once in a while you would sight down the wall to the stakes and see how you were doing."

It has lasted pretty well.

Bill has little good to say about hardheads, those boulders so common in Vermont, worn smooth and round by centuries
of water action.

"They're hard to work with and besides
they weigh a lot. A stone like that will
weigh 180 pounds to the cubic foot, so it
doesn't take a large stone to weigh 200
pounds or more."

Bill would break up many of the larger
rocks to fit them into place. "Working
alone, you had to, unless you could just roll
them into place. You might try for 10 min-
utes to break one, looking for the grain, then
you would tip it another way and it would
break easily." He never swung his 16-
pound sledgehammer against hardheads.
"It's no use. They won't crack. They don't
have any grain."

A man learns something about himself
when building a wall. "Some mornings I
would go out and I would work just as fast
as I could work. And I'd be all in, I'd have
to stop about three o'clock. And the next
day I might go out and just take my time
and work all day and even go out in the
evening. And it didn't seem to matter. In
a day 30 to 35 feet of wall was about what
I could do. You could work fast or you
could take your time and you did about so
much and that's all you did."

He also learned about the enemies of
a wall, the forces that work toward its de-
struction, knowingly or unknowingly.

"Woodchucks will sometimes burrow
under it. Then sometime later the weight

of the stones will collapse the tunnel and the wall will sag a little. And in 50 years the frost has heaved it some, although by now its pretty well settled. The section on the road gets damaged more than the section on the other side near the house. Over the years cars have run off the road and bumped into it. The snow plow has thrown heavy snow and ice against it, and this can damage it.

"A lot of people stop and take pictures of it, and then some of them will sit on it or even walk on it. In the winter the skiers get up and walk the wall. It doesn't help the wall any."

Brush Is Threat   Brush growing around it is another threat. The brush will hide the rocks that fall from the wall, making repairs difficult. And if the brush grows into trees, the roots and trunks of the trees will push at the wall's foundation.

A man's stone wall is his monument.

104

So Burns makes repairs each year, cutting back the vegetation, and replacing loose rocks. "If any have fallen away, I'll pull down a section, then start at the bottom and lay it back up." As Bill shows the wall to visitors, he replaces stones, pulls away vegetation, tidies up the wall, without thinking of what he is doing.

Ask him six times why he built the big wall, and you might get six answers. It's difficult for him to tell a non-wallbuilder why he did it.

"For the exercise." That's one of the quick answers. And he tells how people stopped and asked him to play golf, and couldn't believe him when he said he would rather keep on building.

Better than Golf

A broader reason becomes apparent as he walks beside his wall, checking its familiar lines, admiring its permanence. As a young man he tested and proved his brute strength and endurance against the heavy rocks that make up the wall. Now, a half-century later, his satisfaction is deep as he sees the durable results of his work and lives again in his mind those long-ago challenges of his youth.

Other Storey/Garden Way Publishing Books You Will Enjoy

BUILD YOUR OWN STONE HOUSE: USING THE EASY SLIPFORM METHOD (revised & updated edition), by Karl & Sue Schwenke. Natural, strong, and beautiful stone continues to be a favorite building material in town and country. Highly readable and based on the authors' experience of building their own home, here is the only book on the market that explains the easiest method of building with stone. 160 pages, 6 x 9, illustrations, charts, bibliography, index. Quality paperback, order # 639-8 $21.95 US/$29.95 Canada

BUILD YOUR OWN LOW-COST LOG HOME: UPDATED EDITION, by Roger Hard. Rustic beauty, practicality, and low construction costs. This completely updated classic includes step-by-step instructions and diagrams, 208 pages, 8½ x 11, illustrations, photographs, tables, charts, index. Quality paperback, order # 399-2 $11.95 US/$15.95 Canada

BUILDING SMALL BARNS, SHEDS, AND SHELTERS, by Monte Burch. Construction basics plus plans and easy-to-follow construction methods for attractive outbuildings. 248 pages, 8½ x 11, line drawings, photographs, index. Quality paperback, order # 245-7 $12.95 US/$17.50 Canada

BUILDING WITH STONE, by Charles McRaven. Nothing can rival stone for its beauty and durability. This book will educate the novice and inspire the seasoned artisan with step-by-step instructions for stone walls, buttresses, fireplaces, a home, a barn, and even how to restore existing stone structures. 192 pages, 8½ x 11, photographs, line drawings, index. Quality paperback, order # 550-2 $14.95 US/$19.95 Canada

TIMBER FRAME CONSTRUCTION: ALL ABOUT POST-AND-BEAM BUILD-ING, by Jack Sobon & Roger Schroeder. Highly illustrated how-to book for both beginners and experienced carpenters who want to design and build using this technique. 208 pages, 8½ x 11, line drawings, photographs, index. Quality paperback, order # 365-8 $12.95 US/ $17.50 Canada

BE YOUR OWN HOUSE CONTRACTOR: HOW TO SAVE 25 PERCENT WITH-OUT LIFTING A HAMMER, by Carl Heldmann. Here's the inside story on general contracting — the carefully guarded secrets of the trade as told by a veteran house contractor. Recommended for anyone who wants to be more involved in the building of his or her own home. 144 pages, 6 x 9, sample contracts, charts, construction notes, index. Quality paperback, order # 410-7 $8.95 US/$11.95 Canada

These books are available at bookstores or lawn and garden centers. They may also be ordered from *Storey's Books for Country Living,* Schoolhouse Road, Pownal, Vermont, 05261. Please include $3.25 for postage and handling. Send for our free mailorder catalog.

Index

Basalt, 7
Batterboards, 24
Bird bath (*illus.*), 90-95
   foundation/drainage, 92
   water source, 92-95
Boots, steel-toed, 2

Carts, 16
Chinking, 48-49
Chisel, 20
Cleavage planes, 8
Conglomerate, 7
Corners (*illus.*), 36-39
Crowbar, 16
Culvert drainage, 52-53

Dams and water deflectors, 79-83
   foundation, 81
   shape, 81
   tie-stones, 81
Dams, check (*illus.*), 85-89
   placement, 86-87
   vegetation, 88
Dip, 69
Ditch drainage, 52
Drainage, 52-53; 73, 75
Drilling holes, 62-64
Dust scoop, 63

Ends (of walls), 36
Equipment, 16-21, (*see also* Tools)
   gloves, work, 2
   steel-toed boots, 2
Erosion control, 83-88

Footing, 23-24, 26-28; *see also* Width

Gad-pry bar, 19
Gaps, wall, 57
Gate and latch, 60-65
Gate posts, 61
Geological survey maps, 13-14
Gloves, work, 2
Gneiss, 7
Goggles, 17
Granite, 7
Gravity, 1
Guide cord, 24

Hammers, 20
Hammers, mason's, 17-19
Height/width chart, 39-40
Hinge pins, 64-65
Hoe/pic, 16

Igneous rock, 7, 8-9, 67

Labor, 2-3
Lattice fence, 54
Leftover stone, 88, 90-92
Leverage (*illus.*), 40-44
Lifting, 41
Limestone, 7
Line level, 24-25

Marble, 7; 11
Metamorphic rock, 7; 11

Ordering stone, 66

Pinch bar, 16
Plumbob, 31
Posts, 54

Property lines, 22

Quarries, 13, 15

Retaining walls, 54-56
Rip-rap, 83-84
Rock types (*illus.*), 6; *see also specific names*
Rotten stone, 71
Rubble-filled wall, 45-47
 mortaring, 72-73
 slanting top course, 47
 tying, 45, 47

Sandstone, 7
Schist, 7, 8
Sedimentary rock, 7, 11
Shale, 7, 8
Signing stone walls, 77
Slate, 7, 11
Sources of stone, 12-15
Stairs, 58-60
Stiles, 57-58
Stone
 building, 4-11
 not-so-good, 47-51
 sources, 12-15
 types, 5-7
  igneous, 7
  metamorphic, 7
  sedimentary, 7
Stone boat, 67; *illus.*, 68
Strike, 69

Tie stones (*illus.*), 35, 37-38, 81
Tools (*see also specific name*)

Tools, *cont'd*
 quarrying, 67
 sources, 21
Topo maps (*see* U.S. Geological Survey maps)
Transporting stone, 69-70
Tying, 35, 45, 47

U.S. Geological Survey maps, 13-14

Vegetation, clearing, 23, 76

Walls
 alternate materials, 12
Walls, stone
 appearance, 25-26
 corners (*illus.*), 36-39
 ends, 36
 gaps, 57
 gate and latch, 60-65
 height/width chart, 39-40
 large stones, 31
 laying out, 22-23
 not-so-good stone, 47-51
 selecting stones, 29
 settling, 32, 35
 small stones, 31
 stairs, 58-60
 stiles, 57-58
 tie stones, 35, 37-38 (*illus.*), 81
 vertical fissures, 31-32
 width, 39-40
Water deflectors, *see* Dams and water deflectors
Width, 39-40
Wobble knob, 29